Aquaponics for Beginners

An Aquaponic Gardening Book to Building Your
Own Aquaponics Growing System to Raise Plants
and Fish

By

Garrick Mitchell

Disclaimer

This publication is designed to provide competent and reliable information regarding the subject matter covered. However, the views expressed in this publication are those of the author alone, and should not be taken as expert instruction or professional advice. The reader is responsible for his or her own actions.

The author hereby disclaims any responsibility or liability whatsoever that is incurred from the use or application of the contents of this publication by the

1

purchaser or reader. The purchaser or reader is hereby responsible for his or her own actions.

Table of Contents

Introduction

Aquaponics is the combination of two interesting concepts; aquaculture and hydroponics. In a layman's understanding, aquaponics refers to the bio-integrated system through which two elements are being groomed. These two elements are in a mutually useful relationship where the hydroponic vegetables (one of the two elements) are watered and fed from the waste of the fish's pond and in return, it cleanses these wastes from the aqua system. The vegetables feed naturally from the bacteria from the fish that enters into its roots and this helps them to grow properly and even more effectively than non-hydroponic vegetables. This could also be referred to as gardening and aquaculture – it systemizes continuous crop production, aqua life maintenance and water consumption. The aquaponics system allows you to effectively manage the two systems without anyone exclusively suffering.

There are several types of an aquaponics system. Every one of them has the same function, but with different systems of maintenance or building. Starting an aquaponics system requires you to pay attention to the type of system that perfectly suits you and the environment you wish to establish your garden in. One vital component to running a successful system is the pH – an element that is a highly essential aspect of the aquaculture system, so it is very important to understand the pH system of the water and the fishes when building your own system. A neutral pH of water 6.8 and 7.2 is better for the aquaponics garden. Hence, you have to keep it within the neutral range.

This book discusses virtually all you need to get your aquaponics system up and running successfully from start to finish.

So, let's jump right into the first chapter.

Chapter 1

Essentials of Aquaponic Gardening

What is Aquaponics?

When a bio-integrated system contains two elements: fish or other aquatic cultures also called aquaculture and hydroponically grown vegetables (mostly vegetables), we call it Aquaponics. It is a combination and integration of two subsystems that are biologically cultured. These subsystems integrate through a water medium. These water medium cycles between the fish and vegetables, hence the metamorphosis of aquaponics. Furthermore, in creating aquaponics, there is a 3rd element called the microbial. This is the useful microbial. Microbial could also be a community relating to a microorganism, especially a bacterium causing disease or fermentation. In relation to aquaponics, microbes facilitate the fermentation or nitrification process. The water which contains fish waste supplies the nutrition of the vegetables; basically, nitrogen nitrates. This nutrition is used by vegetables to strengthen their rapid growth. The vegetables after receiving this nutrient take up excess nitrogen and provides water that has been purified back to the container where the culture is occurring. The role of nitrogen in water is purification. Hence, when vegetables release them, it acts as a purifier. Essentially, aquaponics is a mixture of these three living communities (vegetables, aquaculture and microbial) that produces a self-sustaining system that provides plant and animal produce in diversity.

History of Aquaponics

No system exists without history. To further understand the aquaponics system, it's pertinent that history is studied. Studying and understanding the background of aquaponics is important. Aquaponics is fast becoming a system that's sustainable for modern farming. Because even when a farmer has limited space, aquaponics can still be created. In other words, limited space isn't a barrier in aquaponics. History points the practice of aquaponics back to 1000AD. This was a time when the ancient man found that there exists a natural relationship between vegetables and fish. They found that these organisms coexist and luxuriate in one another.

When was aquaponics invented?

It is been debated among scholars concerning the invention and introduction of aquaponics as an agricultural system. It was believed that the primordial Aztecs in Central Mexico started the system of aquaponics around 1000AD. There had to form the system because that that they had limited spaces to farm and grow their food. Thus, that that they had to form a pastoral vehicle that can food crops. The Aztecs had their settlement positioned on the seashores of Creek Lake, Lake Texcoco. This settlement area of the Aztecs had rough soil and thus the land was majorly marshland thus not suitable for growing food. This led to the term "chinampas". Chinampas could also be a garden that's believed to float.

To make the garden, the Aztecs, accumulated decayed vegetables and dirt into Islands that were fixed. The farmers will then plant crops like maize, tomatoes, and pepper on the "Chinampas". The country of origin of aquaponics is typically ascribed to Mexico. Although the origin is additionally largely under debate. Countries like Southern China, Indonesia and Thailand in Eastern Asia were believed to practice aquaponics within the earliest times. Thus, while the Mexicans had their Chinampas People in Southern Asia grew rice in paddy fields with the help of fish. Furthermore, it wasn't unusual to determine farmers in South China raising ducks housed in cages that were placed on top of a fin fish tank. The fin fish successively produces waste that's used to feed another community of cat fish. And thus the waste water from the cat fish was used to water and fertilize the rice fields and other crops. It was a cycle of feeding on the fin fish to the cat fish and right down to the rice field.

About Dr. Mark McMurtry

Talk about a renowned man, then talk about Dr. Mark McMurty. He is known as the father of contemporary aquaponics. He is renowned for creating the first closed circle system of aquaponics, in conjunction with Prof. Doug Sanderson. Dr. Mark McMurty along with other research fellows researched into aquaponics in the Mid-1980s. Their research into aquaponics was inspired by the experiments administered at New Alchemy Institute. In this farming method, wastewater from an aquarium was recycled to irrigate and water crops, such as tomatoes and cucumbers in grow-beds full of soil. The soil (sand) also functioned as bio-filters, which sieved out wastewater that flows through it. The drained water from the sand was recirculated into the aquarium.

How Does Aquaponics Work?

Aquaponics could also be a system that works in a cycle repetitively. What aquaponics does is grow both fish and vegetables utilizing the nutrients that come from both communities. The major merit of aquaponics is its ability to point out the disadvantages of both aquaculture and hydroponics and turn them into an advantage. This is possible because aquaculture (growing fish) demands an outsized body of water, which easily gets saturated with waste and harmful chemicals, thus requiring water change as often as possible which is kind of expensive. In the same vein, vegetables that are grown hydroponically requires artificial fertilizers to grow because pure water is poor in nutrient. It is these two drawbacks that aquaponics utilizes to urge the best results. Aquaculture provides nitrogen to the vegetables which helps the vegetables grow. The wastewater from fish contains nitrogen and other secondary elements in large quantities. The vegetables reciprocally supply oxygenated water to the aquarium.

The Cycle of Aquaponics

The cycle flows from the fish to the vegetables and from the vegetables back to the fish. The fish enrich the water with mineral elements by producing and release excrement into water. The water passes through a biofilter and circulates. A bio filter could also be a colony of nitrifying bacteria which turns ammonia into plant nurturing nitrates, in other words, a bio filter makes the water healthy and suitable by purifying it. After this, vegetables consume the water that contains the excess nitrate and other mineral compounds for its growth. The vegetables then release purified water which is returned to the aquarium. It is this cycle that is called aquaponics and typically produces a high yield of food crops and fish.

As stated at the start of this chapter, microorganisms may be a subsystem that plays an important role in aquaponics. They make the organic process within the aquaponics system work because of the bacteria which break down nitrogen. The microorganisms utilized in aquaponics are useful ones. As you would have possibly noticed, besides aquatic animals and vegetables, there's a 3rd group of organisms that have a key role in aquaponics. These microorganisms also are called biofilters. They assist nitrification by removing the ammonia and nitrate in water that is toxic to fish. Additionally, nitrification provides vegetables with the nitrogen they require for their growth.

Main Living Components of Aquaponics

The three major components or subsystems of an aquaponics system are fish, plant and bacteria.

Fish

This is often the component that supplies nitrogen and other mineral elements to the plant via its wastewater. These nutrients help the plant grow rapidly. It's the aquaculture component within the aquaponics system. To take care of the fish, enough quantity of water must be supplied within the tank. The fish are kept during a tank often translucent. It advised that for each pound of fish, two gallons of water are needed. Water helps to live the general health of the fish. When selecting the fish for aquaponics, they're several factors to think about. One of these factors is that the fish must be ready to survive in a high-density population condition. The fish must be able to survive in an indoor tank. Some fishes cannot survive in an indoor space. This is often why most farmers choose Tilapia. A tilapia can sleep in a high-density population condition and also survive in an indoor space. Other known and commonly used aquaponics fish are catfish, trout, perch and hybrid striped bass. These fish species would be treated in detail in subsequent chapters.

Another thing to think about is that food fed to the fish also called fish feed. Whatever fish food that's chosen must be ready to drift on the water. The essence is so that the fish can easily catch the feed as they swim. Although feeding standard differs based on fish type.

What Fish Are Best For Aquaponics?

Not every fish can be used for aquaponics, even amongst those that can be used, some are more effective than others.

Tilapia

Tilapia are currently the foremost popular and simply raised food fishes. They will survive a highly densely populated living conditions and may survive in an indoor space. Additionally, they mature quickly thus reaching harvest size quickly. They will also easily adapt to water pH levels. This fish is also not carnivores, thus they don't prey on younger fish.

Catfish

These are the leading, commercially raised seafood. Cat fish is suitable for aquaponics because they're not easily susceptible to diseases and parasites that are obtainable in enclosed tanks. However, catfish flourish may be a rarity area.

Goldfish

Cousin to the carp, are a well-liked, non-harvestable choice for home systems. Tough and straightforward to get, goldfish and their larger relative Koi fish, add decorative touches to your system. Both are an honest choice for beginners.

Carp

These are hardy and adaptable to a good range of conditions. They will survive the foremost difficult living conditions starting from temperature, water levels, pH levels among others. Thus, they're perfect for beginners in aquaponics. However, carps aren't the first choice due to their taste which is usually muddy. But when the carp you grow are in clean, aquaponics water, they don't have the muddy taste of carp taken from rivers and lakes.

Trout

There are a favourite fish but are harder to raise especially in aquaponics. They require a highly monitored temperature level. Trout survive during a relatively cold-water temperature which is about 55 degrees or lesser. The water from Trout wouldn't be suitable for the vegetables in an aquaponics system, cold water will hinder the expansion of vegetables. Tomatoes, cucumbers, squash and other warm-season crops aren't fitted to systems of raising trout or other cold-water species.

Tips & Techniques for Fish in Aquaponics

1. To stop a replacement or new fish introducing disease to your aquarium, quarantine for a minimum of 3 to five days.
2. Avoid copper plumbing in your system, it's harmful to fish.
3. If your aim is to watch your fish to survive and grow rapidly, you want to ensure the temperature in your aquarium is stable. Fish are

sensitive to natural processes and any change in their environment can be toxic to their survival.

4. Avoid exposing outdoor fish tanks to the sun which may cause rapid water temperature fluctuations.

5. To stop the rapid climb of algae, provide shade or protect fish tanks.

6. One important thing to note when planning your aquaponics is to calculate the density of fish to water within the tank. Too many fish in a tank can starve the fish of oxygen and otherwise stress them. Too few fish mean less nourishment for your vegetables.

Plants

This is often the component that utilizes the nitrogen that's supplied from the excrement of fish. The plant successively supplies purified oxygenated water to the aquacultured fish. They purify the water by utilizing nitrates through absorption. Thus providing clean water to the fish in the tank. Noteworthy is that the plants often utilized in aquaponics are vegetables. The role of an aquaponics system is to extend food security by providing a sustainable system.

The plants are often grown on a container called grow bed, which holds the nitrate-rich water, and floats the expansion base. The medium used for plants also called grow bed is usually translucent and preferably black. However, opaque containers are considered better because they need a way to reduce the chance of algae explosion in your tank, unlike translucent containers. To carry the plants, a base that is light in weight and buoyant is required. There must even be a net pot that permits water from the plant to the fish.

Recommended Plants for Aquaponics

Although, aquaponics may be a major innovation in agriculture, not every plant are often utilized in aquaponics. The recommended plants comprise root vegetables like carrots and turnips. Vegetable crops like tomatoes, cucumbers, strawberries, watermelon, and pumpkin. Herbs like basil, mint leaf, lettuce and other leafy greens.

Bacteria

This is the useful bacteria that help the purification process of the aquaponics cycle. While they seem not essential to aquaponics, they play an integral role. If bacteria in bio filters are not available, both the fish and plant will lack the nutrients they have for rapid growth.

Through a nitrification process, bacteria in biofilters serve as a go-between for both the tank fish and therefore the plant. The procedure during which nitrogenous biological compounds are transformed into nitrites, then nitrates is called Nitrification. It is the transformation of ammonia to nitrite. Nitrococcus (in aquatic environments) and Nitrobacter (in soil) additionally dissolve the nitrites into nitrates. Once transformed into nitrates, the composites are now in a component that can easily be utilized by the plants (vegetables) whenever they need to feed. In simple terms, it's an organic process that works by changing fish waste, uneaten food and other organic matter gotten from the fish tanks into nutrients vegetables can utilize.

Types of Aquaponics Systems

Media Beds or Growbed

The media-filled bed or flood and drain system is the foremost used aquaponics system. The aquarium's water is raised with the aid of a solid lifting overflow right into a filter, which in turn separates the waste. This is often necessary if you employ an outsized amount of fish. If you have a small amount of fish or a small stocking density, water can be pumped directly into the grow beds. This technique is employed by backyard gardeners because it's a neater process. A media bed is often made with different media. About anything can be planted within the grow beds, just ensure it doesn't get clogged with solids. A media-filled bed is very vulnerable to rusting which suggests that the nitrifying bacteria in your grow bed will die due to zones that are rotting. While media bed is basically easy to line up, it, however, needs high maintenance and not ideal for commercial systems.

Deep Water Culture (DWC)

Deep Water Culture also called Floating or Moving Rafts are styrofoam boards with 2 to 3-inch net cups in them. Underneath the rafts may be a fixed and continuously flowing water mass that's slow but immerses the roots. Since the roots are immersed you'll get to distribute additional oxygen under the rafts. DWC is mostly utilized in commercial aquaponics systems, most of which are modeled after the UVI design (University of Mary Islands). It's a system that produces tilapia and leafy vegetables. It's a highly successful system. This technique is sweet at making small crops like lettuce, basil, okra. An essential advantage of this technique is that it's a stable temperature and allows for constant water flow thus suitable for commercial purposes. However, it is a heavy system that is not ideal and advisable for a roof top. This implies it can't work for backyard purpose.

Nutrient Film Technique (NFT)

This is the system that's popular in hydroponics. The nutrient film technique or just NFT is a system that uses gutter pipes to make a little film of nutrient water to the roots of vegetables. The NFT system is often used for commercial purposes, unlike the media system. The gutters have a horizontal bottom in commercial systems to assist with water movement. The quantity of water and space required for NFT is little, however, it's susceptible to blockage and only suitable for small crops.

Vertical Towers

The vertical tower is employed by those that have small space. It is almost like the vertical NFT system but better. It allows vegetables to grow vertically. However, not every plant grows vertically which is a drawback. The crop's roots are inserted into a bio media, from where water trickles on. Nutrient-rich water furnishes the roots while ammonia is removed by the bio media.

Deep Flow Technique (DFT)

DFT is similar to NFT but using this method requires a higher water level. The gutter pipes are filled 50% with water, just touching rock bottom. Because the roots are going to be submerged within the water, you'll need to provide oxygen to the roots a bit. You do this by installing air stones within the gutter pipes where the water enters. This system requires a small water pump, but a blockage within the channel might destroy crops.

Wicking Beds

A wicking bed is employed to grow root crops and it is not ideal for commercial purposes. It's of two types: recirculating and non- recirculating. Recirculating will take the water heater to the aquarium (sump) to be used again. A non-recirculating system is simpler but won't take the water heater to the aquarium. It is a raised bed with media within the bottom. The media are going to be submerged within the water all the time. Over this media, you'll have to put a weed barrier. On top of this weed barrier, you would like to place your wicking material. This will be the compost soil, just confirm it takes the water up to the roots of your vegetables. The wicking bed can grow potatoes and carrots. Nutrients are delivered by the wicking effects to the bottom of the plant.

Dutch Buckets

This is an irrigation system that also comes with media beds. They are used in growing big crops or fruits. Dutch buckets are small growbeds that have a trickle of water on a timer. There is two inches of water in the rock bottom of the bucket that helps to supply water to the plant. The drained water will return to the sump or fish tank. However, during this system, water might heat up quickly due to the biomass of the grow media.

Benefits of Aquaponic Gardening

Aquaponics provides a myriad of benefits to man and society.

1. It helps reduce food scarcity through its sustainable system. In aquaponics farming, a cycle for a constant supply of fish and vegetables is provided. And this cycle could also be a system that might always produce result if the right steps are put in place

2. It is used to grow both vegetables and fish. One can grow vegetables and rear fish simultaneously.

It's the equivalent of getting the simplest and the best of both worlds. It utilizes the drawbacks of fishing and planting and changes it into a useful farming system. In addition, the system is fully symbiotic. The symbiotic relationship happens where the fish supplies the necessary plant with nutrients for rapid growth while the vegetables supply the fish with oxygenated water.

3. Fast and rapid rate of growth of crops. The expansion and rapid climb rate of the crops are fast thanks to the adequate amount of nutrients that are made available to them. The fish utilized in aquaponics produce wastewater that is highly nutritious. It's this water that's used to water the crops thus producing a high yield.

4. Production of crops and fish through a controlled and closely monitored environment. The myriad of crops and fish is completed during a controlled environment where temperature among other factors are often controlled.

5. Disease-free environment and controlled environment. There aren't any cases of soil-borne

diseases, pests, and infections because the soil isn't utilized within the system. Also before a fish is introduced to a fish tank, it is quarantined for 2 to 3 days to check if it has any disease that might contaminate other fish.

6. Little or no weed build-up. The structures utilized in aquaponics are contained in a highly controlled environment which will not allow weeds to build and because of this, there is no buildup of weeds within the aquaponics system.

7. Minimal garden chores. The garden chores are minimal compared to plain methods of farming. The only tasks required are feeding the fishes and ensuring that the system is functioning effectively.

8. Aquaponics has a highly flexible farming method. The farming method is often practiced anywhere because of the limited space and water required to run it. One can grow an incredible amount of crops employing a very small space. Aquaponics are often administered at the rear of a house,

hence the term aquaponics gardening. Most of what is required is a tank and a grow bed.

9. Highly scalable farming system. The system is scalable and doesn't require plenty of capital to start and run aquaponics farming successfully. One can even start within the backyard and grow into a commercial system within a short time.

10. The use of fertilizers or any other artificial chemical is not a requirement in this system because the wastes given out by the fish have enough nutrients. This makes the food produce from aquaponics absolutely healthy and safe for human consumption. Additionally, since the system doesn't require artificial fertilizers, it's a less expensive means of farming.

11. Crops are purely organic and healthy. Organic produce has been known to be the simplest for human consumption. Organic produce reduces the expansion of diseases peculiar to man. The crops produced are purely organic and thus the fish are free from any harmful or toxic chemicals.

Chapter 2

Terms Used in Aquaponics Gardening

Acidic: It is when the water in an aquaponics system has a pH level of about 7.

Aerobic: Having an abundant supply of air or oxygen

Ammonia: Ammonia serves as a fertilizer and it is also the first by-product of nitrogen-bearing compounds that nourish the vegetables in an aquaponics system.

Anaerobic: A situation where there is a shortage of oxygen within the system

Bacteria: The third element of aquaponics that's responsible for transforming fish waste into nutrients for the vegetables to soak up. It is usually housed in a colony called biofilter.

Bell Siphon: Also referred to as flood and drain. It is employed in ebb and flow aquaponics systems to manage the flow of water. Bell siphon is used in aquaponics to control water flow. The water is sent back into the grow bed and at a selected point either 2 or 3 inches below the surface. The water is drained away from the bell siphon. During the water draining process, it usually produces a water gurgling sound. The Bell Siphon helps water movement between the plant and the fish.

Biofilter: This is where bacteria are kept for colonization. It is usually a mapped-out area. The water pumped from the aquarium runs through the biofilter, where bacteria transformed into ammonia is changed to nitrite then into nitrate again. It is this nitrate that is useable by vegetables.

Biomass: The entire weight or amount of vegetables or animals within the aquaponics system.

Flow Rate: It is the proportion of water a water pump puts out per minute, laid out in gallons per hour (GPH) or gallons per minute (GPM), or it is the speed of flow of water in a pipe in an aquaponics system.

Grow Bed: An appropriate container that's filled with growing media like gravel, grow media, or lava rock. In the grow bed, the vegetables are cultivated and grown. It can also be called a nursery as it is where the seeds of a crop are planted.

Grow Light: Also called "plant light". The Grow Light is a man-made source of sunshine designed to stimulate plant growth by emitting light suitable for photosynthesis.

Grow Media: Grow media may be a soil replacement utilized in aquaponics gardening. Samples of grow media used in an aquaponics system are pebbles from clay, lava rocks, expanded shale, and small gravel stones.

Nitrobacter: A bacteria that change nitrites into nitrates and play a crucial role within the organic process by oxidizing nitrite into nitrate in soil and aquaponics system.

Nitrites: Produced by bacteria and a source of food for Nitrobacter and a product of Nitrosomonas. It produces nitrogen.

Circulating System: A system during which the water isn't diverted away from the system but is recirculated and then reused again. Aquaponics is simply a recirculating and recycling system. Virtually everything is reusable.

Sump Tank: A tank at rock bottom. A subpart of the system. It keeps the aquarium water level constant. Sumps are utilized when the water moves from top to bottom. In this case from the fish to the plant and vice versa. The flowing water is then transported back by a water pump.

System Cycling: Refers to the method of building a bacteria colony that is healthy within an aquaponic system.

Aquaponics Gardening – This is a homegrown aquaponics gardening where the harvest is to be eaten by the gardeners themselves and is not cultivated for commercial resale purposes. It can also be done in a community environment.

Chapter 3

Planning Your Aquaponics System

Size of The System

The size of your aquaponics system remains a factor to contemplate when setting up your system. The size can go a long way to determine if your aquaponics system will thrive or be an utter failure. The size here is not area size but volume. While an aquaponics system does not require a large amount of water, you'll want to make provision for a minimum of a 50-gallon tank for your fish. After all, fish live in water. That is the whole idea of the aquaculture system. Ideally, you should consider starting with an equally larger system. Larger systems are easy to manage and have less fluctuation in temperature and pH levels. Even more important is that the ratio of the dimension between the tank and plant bed must be standardized. According to Sylvie Bernstein, beginners are advised to start with a 1:1 ratio.

The essence of starting with a 1:1 ratio is to ensure standardization of process and reduce the error rate. Another reason Sylvie Bernstein recommends a 1:1 ratio is to ensure the systems are getting to be cycled properly. Once the bio filter is well developed and the farmer has gotten some experience and can handle the process effectively, then he can start increasing the plant bed size. In fact, the farmer can go up to a ratio of 1:4 (tank to plant bed volume). However, at this ratio, the farmer will require a more advanced system with sump tanks and indexing valves. Basically, he has to be certain that the water is getting cycled into the plant bed gradually. Remember that the whole point of aquaponics is the water cycle. Otherwise, he would get an excessive amount of water flowing into the biofilter directly, thus reducing its effectiveness.

Location of The System

The location of an aquaponics farm requires a thorough thinking process. For commercial farming, you need to consider a place situated away from your main living building. However, for a DIY option, think very carefully about where you'll put your DIY aquaponics system. If you don't plan your location thoroughly, it becomes difficult to maneuver. There are options; it could be outdoor or indoor location. If you'll have to move your aquaponics system (such as bringing it indoors during winter), then you'll need to factor this into the planning by adding wheels for easy movement of your system. The reason you don't want your system outside during winter is that an aquaponics system needs a regulated temperature. Winter is a period of extreme coldness and that can lead to the death of your fish.

Location Options

Option 1: Outdoor Systems

The outdoor system is often used by those who reside in places like Australia, Texas, Florida, or Southern California. They have a more balanced temperature throughout the year thus the vegetables and fish can survive outdoor. Unlike places like the UK where the weather is a bit of an extremist. It isn't adequate for the temperature to be warm for all-year plant cultivation. You furthermore may require the temperature to be equally stable so your fish can thrive. So when considering the outdoor location of your system, consider the survival of both your plant and fish. You are not advised to grow your system seasonally. It should be all-year-round because of your biofilter – when your biofilter gets going, hardly would you get bountiful harvests from the system.

Option 2: Bringing System Indoors During Winter

This is the option of having to move your system from outdoor to indoor because of winter. Of course, before this, you must have designed your system to be able to move on wheels. Given that growing seasonally is practically a wasted effort, you would possibly consider bringing your outdoor aquaponics system indoors during the cold months. This gives you the liberty of effectively regulating the temperature of your system. You can simply structure and strategize to build your system ahead of the winter by providing wheels, so it's easily moveable. This strategy works well. Although, you'll require grow lights and some equipment to adjust and create a stable temperature.

Option 3: Indoor Systems

An indoor system is used if you aim to run an aquaponics farm for personal consumption; like indoor gardening. Indoor gardening system is straightforward and affordable thus suitable for systems intended for family consumption. However, once you begin delving into larger systems because you realized how effective an aquaponic system is, you'll encounter problems with your indoor system. Some of such problems include; growth of mold if not properly controlled, because grow lights need power and produce heats. An aquaponics system is heavy. There is the tank and grow bed. You have to confirm your floor can support the load. Water spills are inevitable and can lead to algae growth which will cause damage to your system. It will require consistent monitoring to ensure the temperature is regulated. You have to watch out for fish disease anytime you may require to introduce a new fish to the tank.

Option 4: Greenhouse

This is the perfect location solution for aquaponics. Greenhouse is the solution for cultivating all-year-round. The greenhouse system provides a stable environment for both the plants and fish to flourish. The water serves as a heat sink and provides the humidity necessary for greenhouses. In addition, a greenhouse location is perfect for commercial aquaponics farming. Although you have to determine whether you've got the space for a greenhouse on your property. You'll have to construct the greenhouse which will cost you some money, get appropriate permits from the governments, and prepare for water and power requirements.

Fish Stocking

Stocking density may be a major consideration for anyone starting out in aquaponics gardening. As numerous people have asked about the simplest density for crop production, Dr. Nate Storey has given us an overview of stocking density and a few great pieces of advice that apply to starting, maintaining and managing a healthy and productive aquaponics system.

Aquaponics Stocking Density

The amount of fish to start with in your tank is dependent on the species of fish and its feed requirement. However, there are certain common procedures you'll follow when planning and strategizing your system. The recommended volume is 1lb of fish each square base/foot of grow bed area and 1lb of fish each 5¬10 gallons of tank capacity. Recall that you intend to help your fish to grow rapidly. Begin cultivation with fewer vegetables when the fish are young, and add vegetables as soon as the fish grows bigger.

Recommended Rule of Fish Stocking Density

Remember that each aquaponics system has factors and principles that have got to be considered when making choices about the type of fish you decide on and therefore the density it will support. The general guideline often followed by aquaponics farmers especially for beginners is to choose lower density. If you want to experience less stress then you have to opt for lower density. Greater densities require more complicated filtration and aeration setups. Lower density keeps things simple and easy for you to manage. This is often the recommended target. You can use this system: a pound of fish for each 8-10 gallons of water. At the maturity of fish, you can increase to one pound per 8-10 gallons as your system thrives and develops. This is the best place to start out and it could even work as a rule, but make certain to pay close attention and assess the water requirement in your system consistently. Monitoring the system will assist you to form adjustments appropriate for your production goals. The blueprint to a successful aquaponics farming is close and proper monitoring. The moment you start planning, it'll save tons of your time and trouble to think about how your system will function and figure the stocking density accordingly. Do not forget that these guidelines are general.

Feeding Rates

When you are thinking about setting up an aquaponics system, the fish feed is one of the foremost important things to consider. There's an immediate relationship between the feed going into the system and the rapid development of your vegetables. It is a symbiotic relationship. You add the feed into the system, the fish eat the feed and produce ammonia, bacteria change ammonia into nitrates, and the vegetables consume the nitrates and grow. Due to this relationship, balancing the right amount of feed entering the system is important to the general success of the system. When sizing an aquaponics system for feeding, you can use this prescription; 20g of feed for each square meter of the culture growing area. Using this formula, you'll calculate the quantity of feed necessary in your system. It is imperative to follow a standard estimate to avoid reducing the efficiency of your system. Furthermore, when feeding, only give the fish the food they can consume in 5 minutes – remove any unconsumed fish food to avoid undesirable water quality issues. If your aquaponics fish don't seem to be hungry at the moment, don't feed them. Fish can stay for weeks without feeding. The best you can do is to monitor them closely in order to tell the time they ought to be fed.

In a situation when you are cycling with fish, feeding should be kept to the barest minimum during the major 1-2 months until your nitrifying bacteria can easily and quickly change ammonia to nitrites. You can increase the level of the feeding of your fish once the nitrifying bacteria can easily utilize ammonia and nitrites in your system. Again you have to closely monitor and observe your fish in five minutes to see what proportion the fish consume. Some farmers feed their fish 2-3 times each day to foster a gentle metabolic digestion rate and maintain a gentle amount of solids.

Biological Filters

Many aquaponics beginners are often confused by the necessity for filtration or a biofilter in their aquaponics system. Many growers want to be sure that their system requires a biofilter. The major reason for aquaponics is that it's possible to duplicate nature and create a suitable growing environment for the crops and fish to grow. Aquaponics is generally a recirculating method of farming that mixes aquaculture and hydroponics (two different subsystems), resulting in the raising of fish and growing of vegetables in a symbiotic environment with useful bacteria. The sole role of this useful bacteria is to change wastes products from the fish into nitrates, which changes the vegetables' food, and reciprocally, the crop roots filter which ensures the water is purified for fish utilization. A biofilter helps this process to ensure nothing goes to waste. Ordinarily, the wastewater produced by the fish is ammonia which is harmful to the vegetables, but the useful bacteria changes the ammonia to nitrates.

What then is a Biofilter?

A biofilter is a region mapped out for the bacteria to colonize. In other words, it is a colony of bacteria. It makes available large surface areas, pH, dissolved oxygen concentration, and proper temperature. Biofilters are quite simple to line up, and that they contain a tank connected to the vacuum water pump and a few substrates that provide as much as possible, an enormous area for the nitrifying bacteria to grow. The larger the surface area for the bacteria to develop, the higher the chances of nitrification, because it means a more efficient nitrification process.

What is Biofiltration?

Biofiltration is the procedure of changing ammonia and nitrite into a useful chemical called nitrates by the useful bacteria. Ammonia and nitrites are poisonous to aquaponics, even at little concentrations. Basically, an essential requirement for vegetables to grow is nitrate, so a biofilter must be connected to deal with most of the living bacteria. The water circulation within a biofilter will break all the fine solids not caught by the clarifier, thus preventing the waste formed from destroying the vegetable roots within the aquaponics system. Quality biofiltration is additionally indispensable for chemical regulation that occurs in an aquaponics system. Although the case may be different if what you want to build is a media-based aquaponics system. In this case, building a different biofilter is pointless because the grow beds acts as biofilters in a media system. In aquaponics, if the space mapped out in your grow media isn't large and sufficient for the bacteria to inhabit and colonize, you would like to consider a larger space area. Thus the need for a biofilter. This biofilter will become a crucial aspect of your structure by guaranteeing that vegetables have sufficient nutrients to grow healthily and rapidly while purifying the water for the fish to utilize. Your entire aquaponics system may be a biofilter.

On the other hand, if you're using the DWC, Vertical, and NFT aquaponics system to cultivate your vegetables, you'll need a biofilter. In these sorts of aquaponics systems, without a biofilter, the action of bacterial conversion will be low and slow due to a scarcity of adequate area. A raft or media-based system doesn't generally require a separate biofilter as a prerequisite since the raft, growing media (clay pebble, gravel, or lava rock), tank walls, and other surface zones provide enough capacity for the bacteria to inhabit and colonize.

The procedure of biofiltration in aquaponics happens in three main steps. This procedure is for typical aquaponics. The method may vary, which is highly dependent on your biofilter design.

1. The first step occurs when the vacuum water pump pushes the water out of the aquarium and into the biofilter.
2. Once the water enters the biofilter, the water undergoes the nitrification process. This is the process where the useful bacteria help to change ammonia and nitrites into nitrates.
3. When the water filled with nutrients runs through the biofilter and into the vegetation, the vegetable roots absorb the nutrients while purifying the water before it is returned to the aquarium. The water that enters into the

aquarium is oxygenated water that in turn nourishes the fish.

Types of biofilter to use in an Aquaponics system

You can use any of the different filters sold in the market for your system. The type of filter you use is dependent on the type of aquaponics system you are planning to build.

Drip Filter

This filter lets water descend from the highest level. The water flows over the filter box, filled with grow media such as clay, pellets, or other related bio media.

It is important to position the biofilter where the water is purified before it reaches the vegetation. Because when the water flows over the filter, the ammonia is changed to nitrates before it is pushed back to the vegetation.

Static Filter

Static filters are fashioned in a tray-like manner which will be glided into a distinct vessel next to the aquarium before the vegetables. These provide an outsized space for the bacteria to gather on. This filter must be situated next to the solid filter before it spreads to the vegetables, thus allowing the water to be purified before it reaches the vegetation.

Moving Bed Filter

This biofilter is employed during a floating raft system where there's not much exterior space for the bacteria to inhabit.

Benefits of Using Biofilter in Aquaponics

1. Using a biofilter supports the ventilation (aeration) and nitrification processes within the DWC, NFT, and Vertical aquaponics systems.
2. Facilitates and aids the chemical and nitrification stability of your aquaponics system.
3. Absolutely stress-free to use and need little or no maintenance.
4. A good biofilter will lead to the rapid growth of your vegetables by changing ammonia to useful nitrates

Important factors to consider before using a biofilter

Before you opt to use a biofilter in your aquaponics system, you should remember some of these important factors and considerations.

Fish Density

If you're running a manageable size of fish concentration, then you do not need a biofilter. Small quantities of fish won't yield sufficient waste to validate the utilization of a biofilter. The vegetable's roots will supply enough area for the bacteria to change ammonia into nitrates. But in a large quantity of fish density, a biofilter is a must-have.

Remove the Solids

Principally, solids must be separated because this may probably clog the pipes. Once these solids gather in your biofilter, this may create areas often referred to as "aerobic zones" which will entice bacteria but decrease your biofilter's area. This may lessen your biofilter's productivity because it can result in a high ammonia level, which is toxic for both fish and plants. That's why it's imperative that you simply add a typical filter that will eliminate the solid.

Provide Aeration for Bacteria

In an aquaponics system, make sure the bacteria are visible to the air. Using a biofilter in your aquaponics system will enclose the bacteria within the filter. So you've got to be absolutely sure your bacteria have access to air by giving your filter enough air supply. Lack of air equals a lack of efficiency for your bacteria.

Cycling

Cycling is an important concept in an aquaponics system. The question is, how does one cycle an aquaponics system? Nitrogen cycling takes place in the soil and water all around the world. Aquaponics cycling is the natural action of bacteria oxidizing ammonia into nitrates.

Cycling Process

When ammonia is found in your aquaponics system then the cycling process can start. Ammonia may be a compound made from nitrogen and hydrogen. Fish produce ammonia from their feces and urine, but mostly from respiration within the gills. Even as our human waste is harmful to humans, ammonia is harmful to fish and may wipe them out. Of all the nutrients recommended for vegetable absorption, ammonia isn't one of them. The cycling procedure changes ammonia to a less toxic sort of nitrogen (nitrate) that vegetables can easily absorb for their growth. Ammonia attracts nitrosomonas, the primary of the 2 types of nitrification bacteria which will populate the surfaces of an aquaponics system. The nitrosomonas bacteria change the ammonia into nitrites (NO2). This is often the primary step within the aquaponics cycling process.

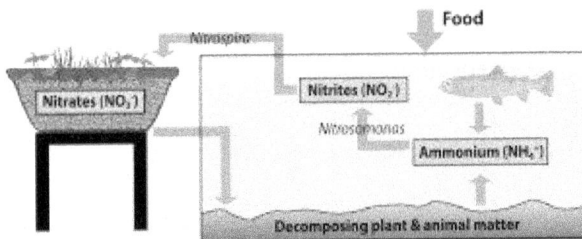

Regrettably, nitrites are even extra harmful compared to ammonia. Thus the cycling procedure doesn't end with nitrites. Nitrites entice the species nitrospira. Nitrospira changes the nitrites into nitrates which can now be utilized by the vegetables. Nitrates are not toxic to the fish and it is an important nutrient for your vegetables.

Cycling Startup

It occurs immediately you start fixing or restarting an aquaponics system. The cycling procedure commonly takes from four to six weeks depending on the water temperature. Water must be stable else the system will take longer to cycle since the bacteria are going to be sluggish to eat and reproduce. This is often a temperature range that's beneficial for both fish and vegetation. You can increase the cycling procedure by bringing a nitrifying bacteria rather than expecting them to thrive independently which can be quite slow.

By adding a bacteria supplement to the media surface or water, cycling happens faster and your system populates rapidly.

Fish Cycling

You can cycle your aquaponics system by introducing fish to the water. The fish will make available the ammonia source from their waste water. When you want to start feeding your fish, you must follow a feeding regimen. For example, it is advised that you don't feed your fish for the first 24hours. You should feed your fish with little feed or feed lightly for the primary number of days. You can include a bacteria starter supplement or allow the bacteria to colonize independently. It is advisable that you conduct a water test a day to see the components of ammonia, nitrite, and nitrate levels.

Required component levels

- Ammonia - below 3.0 ppm.
- Nitrites –below 1.0 ppm.

Nitrates will surge over time. If ammonia goes above 3.0ppm or nitrites above 1.0ppm, then you should remove and replace one-third of the tank water. If you are not careful, the fish can undergo pain and even die if ammonia levels rise above these. Fish cycling is often demanding.

A substitute is fishless cycling which happens by simply introducing a bacteria starter or allow bacteria to move in the surface area naturally and independently. Bring in some amount of ammonia (liquid or powder) until levels are at 4.0 ppm. Carry out water trials to detect ammonia, nitrite and nitrate levels. Continuously add ammonia to keep the level of ammonia around 4.0 ppm. Cycling is done when the level of ammonia and nitrite falls below 0.5 ppm within 24 hours. The existence of nitrates signifies cycling is happening.

Common Cycling Mistake

Sometimes, mistakes occur in cycling. A common mistake is to feature bacteria starter and ammonia only in the beginning. When a water assessment shows 0 to 0.5ppm ammonia, it's presumed the system is cycled.

Categorically this suggests it's time to feature more ammonia. The ammonia must be constantly available for the bacteria to survive.

Another error is to directly introduce ammonia to your fish. Just a single source of ammonia is important, and fish yield ammonia independently therefore it is unnecessary to include ammonia supplements within the fish tank. Remember that ammonia is harmful to the well-being of your fish.

How to know if a cycle is complete

When a water test identifies nitrate levels growing increasingly, you will know cycling is done and complete. At an equivalent period, you will notice that ammonia and nitrite are consistently lower than .5 ppm or a lesser amount. It simply shows your aquaponics system is going to be completely cycled thus creating a safe water for your fish and nitrogen for your vegetation. As soon as the nitrification cycle begins, it'll carry on till the source of ammonia is eliminated or the temperature of your system will remain highly unstable.

Nitrification

Nitrification has been mentioned in bits and pieces in this work. To further understand the aquaponics system, you must know that nitrification is the procedure that runs most aquaponics systems and makes them effective. Basically, nitrification transforms ammonia and ammonium into usable nitrate, which occurs in two phases: transforming ammonia into nitrite, and transforming nitrite into nitrate. In most ecosystems (excluding anaerobic) ammonia is rapidly transformed into nitrite ($NO2-$). This is done by microbes which in the presence of oxygen oxidize and changes ammonia. The major microbe that does this job is Nitrosomonas. A recent study reveals that there are hundreds of numerous, if not thousands of various species (including Nitrosomonas) that does this transformation.

Changing to nitrite is the first stage in the nitrification process. The next stage within the cycle is to change that nitrite into nitrate. Nitrite is additionally quite poisonous thus you have to avoid an excessive amount of it in your aquaponics system. These bacteria break down the nitrite and expend energy to produce a chemical called nitrate ($NO3-$). Nitrate may be a moderately non-toxic sort of nitrogen that vegetables can take up and use to create cells. The bacterium that has been most ordinarily known for carrying out this reaction is named Nitrobacter. Studies indicate that many other bacteria contribute to this reaction asides Nitrobacter. As the bacteria oxidize ammonia and nitrite, they release hydronium ions into the tank, making the system more acidic. (For people that want to grow their systems within the recommended pH range for nutrient availability, nitrification is just the distinct best vital procedure for reducing pH). It reduces the acidic pH level. Noteworthy is that nitrifying bacteria are commonly unproductive when it involves changing system variables. They habitually die out or go inactive when exposed to an excessive amount of light, temperature instabilities, instabilities in salinity and pH. The known balance between pH and nitrification effectiveness and productivity has been buttressed by the idea that nitrification activity in aquaponics systems was predominantly a function of two diverse groups of bacteria: Nitrosomonas spp., and Nitrobacter spp. These bacteria are basically responsible for the conversion process.

Nitrate

The top product is nitrate (NO3-). Some vegetables can take up ammonium and utilize it. However, most have a preference for nitrate. When a system has more ammonium, the vegetables are likely to be often less marketable and attractive. On the other hand, in systems with many nitrates, the issue with pests especially aphids are often more intense, thus demanding more attention. It is pertinent that you remember that systems with too many nitrates concentrations would have incessant trouble with pests. Nitrate, nitrite, and ammonia levels are often confirmed easily with a freshwater test kit. Nitrate dissolves within the solution and is immediately contested for by bacteria, fungi, algae and other vegetables. All of those organisms are taking nitrate up and utilizing it in building their tissues. Because the bacteria, fungi, and algae die, once nitrogen (often in the form of protein) recycles back into the system, the cycle begins again. Much of the nitrate, however, is transported successfully to the basis zone, where the vegetables in the aquaponics system take it up and utilize it to grow rapidly.

What is an ideal nitrate level?

It is imperative to calculate the ideal nitrate level. While it's dangerous for the ammonia or nitrite levels to be more than 2 ppm and 1 ppm respectively, nitrate exists above 100 ppm (well off the chart for several nitrate tests) without being toxic to your fish. Many hydroponic systems run nitrate within the range of 160 ppm. Vegetables can often accept levels even above that, but the aquaponics farmer must strike an equilibrium between the requirements of the fish, the system ecology (including pests), and the essential needs of the vegetables. For this reason, it is highly recommended that the majority of aquaponics growers pay careful attention to ensuring their nitrate falls within the range of 40-80 ppm for consistent and rapid plant growth.

How do you maintain stable nitrate ranges?

It can be an arduous task sustaining and maintaining a standard nitrogen level, especially because the system matures, vegetables get large and therefore the system ecology becomes more multifaceted. This might necessitate feeding to be enlarged to satisfy the bigger demand. Most people would immediately want to extend stocking density, which is an error. The best option is to increase feeding rates and at the same time ensure the feeding is not overboard, and see if greater nitrate levels are often achieved with an equivalent quantity of fish.

Pest Control

You might be wondering if pest disturbs a system as controlled as an aquaponics system. The answer is yes. An aquaponics system is not free from pests. Most aquaponics gardeners face plant-munching pests in their aquaponics vegetables from time to time, and typically, these aren't that big of a deal. However, sometimes pest numbers rise to a threatening level which can cause damages to the vegetables. As many aquaponics growers come to be more and more conscious of the probable destruction caused by exposure to synthetic chemicals, many desire to avoid this method and utilize other pest control approaches. Since an aquaponics system is reputed for being a natural system devoid of chemicals, most farmers avoid using chemicals to control pests. So far, the foremost beneficial and current technique in regulating pests in aquaponics is by inhibiting them from eating your vegetables and bore holes. If you did not inhibit pests from entering your garden, you will regulate them by categorizing the pests and using strategies in managing them without the utilization of chemicals.

Pests can cause some unimaginable destruction to your vegetables. But to unravel the pest challenge, we must recognize these pests first.

Some commonly known pests that are often in aquaponics gardens are:

Caterpillars: In small quantities, caterpillars won't be a drag to your vegetables, however, if left unrestricted, caterpillars can destroy crops, comprising vegetables and decorative vegetables during the late spring through fall. A large number of caterpillars can usually cause great harm to vegetables by sucking plant leaves. These are the worms of moths and butterflies

Control Options

1. Squishing: Always look around the rock bottom of your plant leaves and appearance for the little yellow eggs and take away them before they hatch. These are the eggs of a caterpillar. Regular inspections of your garden will help keep the destructive caterpillars far from your system.

2. Exclusion Nets: This where you'll put all of your caterpillars and canopy them with a protective net. This method also will shield your vegetables

away from possums. Using this method will work perfectly.

3. Organic Spray: Organic aerosol can also be applied. You can try using a biological spray to manage the destructive caterpillars in your system. You can ask your farming dealer for a safe-to-use biological spray that will not cause harm to the other insects in your garden.

Potato worm in vegetables: This is also called the tomato hornworm or hornworm. It is one of the foremost negative pests of tomato, pepper, potato, and egg vegetables in any orchard. They're going to devour leaves, little stems, and masticate fragments from fruits, and notwithstanding their enormous size, hornworms are often complicated to spot thanks to their defensive coloring. Gardeners often discover the large areas where the feeding and damage occurred before they see these pests.

Control Options

1. Handpicking: Hornworms are controlled by handpicking them. They are easily seen between dusk and dawn. They tend to feed on the surface

areas of the vegetables. Because hornworms are huge, most gardeners regulate them through handpicking. Once picked from the vegetables, they're killed with soapy water. This is done by dropping the hornworms in a soapy water.

Aphids: Aphids are little like insects that bore and suck the juice from vegetable leaf, stem and twigs. Often called vegetable lice. While aphids don't bring about substantial damage immediately, they spread quickly, damaging crops and spreading diseases. Aphids suckle on plant juice and are detrimental to plants because they carry plant diseases with them. And since they reproduce and breed quickly, they're difficult to expel and control on a farm. So it is vital that as soon as you see them growing in your garden, you should try to regulate them immediately.

Control Options

To control Aphids, you can remove them physically. This is often appropriate for insignificant aphid invasions, where it's possible to urge obviate the aphids from the vegetables physically. All you need to do is wear a pair of gloves you use for gardening to protect yourself and pinch the pests from the stems and leaves.

You'll also prune and cut off the stalks or branches infested by the pests especially if the invasion is simply confined to a limited number of stalks or branches. Then put the trimmed stalks or branches into soapy water for the aphids to die.

MealyBugs: Mealybugs feed by implanting their elongated sucking mouthparts into vegetables and drawing sap out of the tissue. When unrestrained, mealybugs can bring about leaf yellowing and curling. They're usually found in warmer climates. To identify mealybugs, they are wingless insects, soft-bodied that always look like cotton mass that is whitish on leaves, stems, fruits and vegetables.

Control Options

Soft bodies, light mealybugs invasions are often regulated by applying a Q-tip dipped in lotion on the insects. Useful insects occasionally can prevent or control the expansion of mealybugs. Ladybug/beetles, hover fly larvae, and green lacewings are well-known useful insects. To attract useful insects, you can plant herbs like garlic, clover, dill, mint, catnip, and oregano in your aquaponics. You can also encourage wasps' colonies close to your aquaponics gardens.

Using useful insects that prey on vegetable-damaging pests in your aquaponics garden is the simplest technique of natural pest control. Several insect varieties are often utilized in aquaponics, and you will naturally attract these useful insects by building an environment for them in your aquaponics garden. These useful insects can also be purchased from a garden or farming supplier near you.

A Short message from the Author:

Hey, I hope you are enjoying the book? I would love to hear your thoughts!

Many readers do not know how hard reviews are to come by and how much they help an author.

I would be incredibly grateful if you could take just 60 seconds to write a short review on Amazon, even if it is a few sentences!

>> Click here to leave a quick review

Thanks for the time taken to share your thoughts!

Chapter 4

Setting Up Your Aquaponics System

The media-filled bed or flood and drain system is the most recommended aquaponics system for beginners. So, this section focuses on setting up a DIY media bed for your aquaponics system.

Tools and Materials needed:

- Miter Chop-Saw
- Power Drill
- Hole Saw drilling bit
- Utility Snips
- 50 Gallon Aquarium Tank –
- Water pump (550 GPH)
- Freshwater Aquarium Water Test Kit
- Grow Media
- Two Grow Light
- Storing Vessels (3)
- Bolts - 3" & 1"
- Plywood - 48"L x 21"W x 0.75"H 1x4
- Sheets - 15" L (16) 2x4

- Nails - 64"L (2), 58½"L (4), 47"L (4), 20"L (2), 17¼"L (2), 15"L
- (6) 4x4 Poles - 12"L
- (6) Protective equipment (PPE, i.e. hearing protection, and eye protection whenever using power tools and even hand tools)

How Much Does It Cost?

The exact amount to start aquaponics farming is not fixed. Although I have written 350USD. This is because sizes differ and the choice of tools to use differs as well. The choice of tools you will use will depending on your farming goals. Some aquaponics systems setups were built at zero cost from recovered and recycled materials. However, a minimum requirement will be the need to buy a water pump. And, unless you intend to go fishing yourself, you will be required to buy fish.

The System includes the container, grow beds, media such as pebble, gravel, clay, and plumbing lights. Fingerlings, seedlings, testing equipment feed, power, and water. If you're working on a low budget, then I suggest this: start with a little system you can build yourself. Use low-cost fish like goldfish for the first time.

With a small system, you'll significantly lessen the costs accompanying trial and error. While aquaponics is straightforward, there is a learning curve that comes from first-hand practice. It can quickly become frustrating and expensive once you lose a whole populace of fish just because you couldn't determine how to effectively control pH successfully. By starting small, you save yourself these inevitable losses.

Steps in Setting Up an Aquaponics System

Step 1: Gather tools. Gather the necessary tools mentioned above. Additionally, it is highly recommended that you organize your tools and materials in whichever way works for you so everything is easy to access once you need it.

Step 2: Bring together the base layout 2 of the 4x4 poles. Cut at 12" on a plane surface. With a 4, 3" bolts secure a 64" 2x4 flush with the surface top of the poles (1 on each end of 2x4). Next add a 3rd 4x4 poles on the center of the 2x4 flush to top (roughly 31¼" from each end).

Repeat this for the other side of the base. Attach legs using 20" 2x4's on the outside of the ends just built (4x4 poles facing inwards). Lay plywood board on topmost of the frame on center. Using 1" bolts, secure plywood every 12 inches approximately. For added support, attach a 17¼" 2x4 underneath the bottom surface on all sides of the middle poles securing with 3" bolts (2 from outside of the frame and a couple more for support in the center poles) on all sides of both supports. Additionally, I estimated the size of my aquarium to make enough room for placement.

Then, add 2 anchor supports to the base surface using 15" 2x4's approximately 5" from ends and 3½" from sides (about [4] 3" bolts should do anchoring straight through to the 4x4 poles).

Step 3: Assemble grow bed frame layout [2] 47" 2x4's laying [2] 58½" 2x4's across. Leave about 1½" on each end of horizontal 2x4's and secure 1 flush to top using [3] 3" bolts on each end. Estimate down 11½" from the highest of the 47" 2x4 and secure second 58½" 2x4 (still

leaving the 1½" space on the ends). Afterward, link the edges together using the leftover 15" 2x4's, padding them up with the horizontal 2x4's and securing using [2] 3" bolts on each end. To reduce weight for transportation, postpone the shelving until the entire unit is completely arranged.

Step 4: Fasten the body frame to the base. It is better to leave the two separated for a better approach to moving the unit particularly if you're moving by yourself. You can set up the grow bed frame on top of it and then move the vertical poles on the surface of the anchors you kept for support on the bottommost surface. Using [2] 3" bolts on every pole, secure the bottommost of vertical poles to the support anchors from the surface in the assembled unit.

Step 5: Place shelving. Place 1x4″ boards evenly amongst the two shelf frames. Secure boards using 1″ bolt/nail on each of the ends.

Step 6: Assemble grow beds. With a ½″ hole saw drilling bit, drill a hole in one of the ends in both grow beds. Use a pointy razor blade to smoothen the rough edges from the bored hole exit. You can take out large burrs or widen holes if essential. Then using utility snips to secure the bulkhead to the container while adhering to the directions given on the package. The next thing is to cut out your pipe using ¾ PVC. Cut out 3 pipes for every grow bed.

Slide the 3 cut pipes on the opening height. This height will be what will ascertain the maximum water level of the grow beds. It is therefore imperative that you use an accurate estimate and container. Next, place the grow beds on the highest of the unit then bring them into line under the bulkheads so they can be placed in between the shelf panels. Using an equivalent process, add another bulkhead to a 3rd vessel then place it on the second shelf. This may perform the job of a sump tank to assist in preserving the water equilibrium level within the aquarium.

Step 7: Building your bell siphon. You can now start building a bell siphon. A bell siphon is usually a key constituent to the planning of a flood and drain (or media-filled) system.

In a flood and drain system, the water is filled until it reaches the maximum level which is about the peak of the drain pipe attached to the bulkhead in each grow bed. Next, with the assistance of a bell siphon, remove the water. The water is removed faster and then filled until the air gets to the rock bottommost of the siphon and breaks the suction. You can regulate the frequency and speed of the flood and drain with timers and valves. This is done to help the fertilizer-producing process and water purification.

Cut a number of four (4) PVC which can be marginally be above the height of the grow bed. You can either bore holes round the pipe to a good water drainage system. However, bore holes sufficiently small to sort any large fragments. This segment will serve as a protective screen for the grow media separating it from the siphon.

Cut an estimate of 1¼" PVC to a height around ½" which should slightly be higher than the drain pipe. Fasten a 1¼" slip-on plug to at least one side of the PVC. On the opposite side of the 1¼" pipe, bore holes round the lowest only, to get to 1½" from the base. Bore a ¼" hole on one end of the highest of 1¼" PVC. Then you have to try inserting a typical tube into the opening area of the PVC while pushing in just ¼" of the tube. The tube should just drop straight down naturally, else an elastic band would just do.

Cut the tube to an extent less than the PVC. Bore holes using a driller. This will facilitate the process of air intake to interrupt the siphon once the water has finished draining. Then slide the newly made piece over the drain pipe in grow bed. Drop four PVC screens over the whole drain unit.

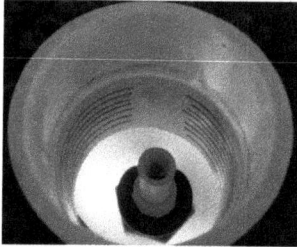

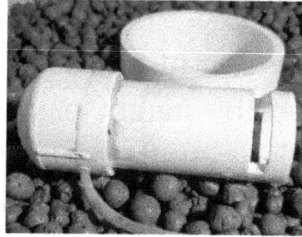

Step 8: Laying out of plumbing. With a different ¾" PVC adapter, work through down to the bottom of the grow beds and the tank for the sump. Fasten a ¾" PVC. Cut out as size 3" on the lowest part of the grow beds.

Drop about 90° length of PVC on the edge of the grow beds facing towards the middle of the sump tank below. Add an additional ¾" PVC. This time, cut at 8". Then try arranging your PVC towards the middle of the sump tank. Place another 90° PVC edge at the top of your tank. Place it facing downward to the angle where water will flow more precisely. Slip a ¾" PVC pipe cut at length in the water equilibrium level of the aquarium from the sump tank.

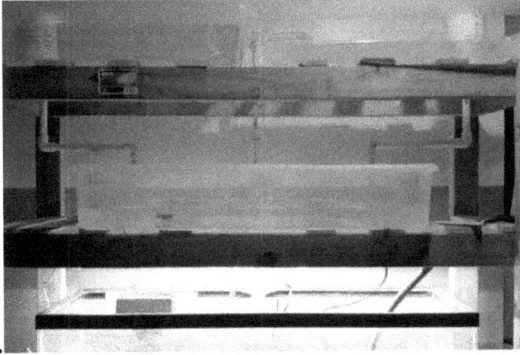

Step 9: Securing the lighting. You can secure the aquarium light by using lighting hardware. Aquarium starts by securing chains to 'S' hooks connected to light. Wrap this chain around each shelf panel at each end connecting back to the hook. Next, drill two holes at the edge of the support frame encompassing the sunshine. Using a 3 bolt pulley, drill the pulley frame to an open beam (or improvised fixture). Finally secure grow light to the pulley ropes.

Step 10: Configuration of the water pump. Sometimes people use a 300 GPH water pump with a 5 feet lift. However, the water often barely flows to that level of height and would fill unevenly. But you can get a water pump that runs 550 GPH. This size of the pump helps fills the grow beds with water faster. Place water pump on the brink of the center of the aquarium. Using a one and half water tube, connect to a water pump running the opposite end to the highest of the grow beds. Then divide the water tube with one tube running to every grow bed. Use bolts to drill two hose clamps to the rear of the unit frame. Utilize these clamps to lock a stick perpendicularly to clamp the water hose. Fill the aquarium slightly above the water pump to allow the water pump to be fully submerged. Once grow beds begin to fill with water, plug in a water pump allowing the water to circulate throughout the whole system numerous times. Scrutinize all linkages and connections for leaks or potential weak spots. Make modifications to the bell siphon to manage water concentration levels if necessary. Adjust water pump rapidity to manage how briskly or how slow the grow beds fill with water. Make any maintenances if necessary.

Step 11: Conducting leak water test. To check for leaks, top up the fish tank to the brim. Top up to a level that is a little higher than the water pump, allowing the water pump to be fully immersed. Once you notice that the grow beds are starting to fill, plug in the water pump to allow the water to circulate throughout the whole system several times. Examine all connecting joints and connections for water leaks or potential fragile spots. Make modifications to the bell siphon to manage water levels. Adjust water pump speed to regulate how fast or how slow the grow beds fill with water. Make any maintenances if necessary.

Step 12: Add uncontaminated grow media. Choosing the grow media are often based upon what you propose to grow. Some media are going to be more suitable to root systems whether or not they need more or less support so as to prosper. For the grow media you choose to select, you may have to put into consideration certain things like pH levels. Certain media are either more acidic or alkaline which is a necessity as a stabilizer of the system. You can also exploit and utilize clay pellets because they contain an almost balanced pH level thereby making it less difficult to adapt to the environment. Plus, the absorbent surface retains more water, keeping the roots moist for an extended time so that they don't dry out. Adhere to the instructions on the packaging and make certain to cleanse the clay pellets methodically to get rid of any dust particles or other debris which will have been accumulated in the process of packaging and transportation.

Add media until slightly less than the highest of the grow bed, based on the water concentration level height set by the bell siphon drain. This could offer you roughly 2″ of dry media on the highest of the beds in the least time.

Step 13: Cycling your aquaponics system. To determine the effectiveness of your aquaponics system, it's essential to cycle the entire system. Cycling often happens between 10 days up to 5+ weeks. To start cycling, you would want to try to run your system with an ammonia presence for a couple of weeks until the system begins producing nitrates (this has been extensively discussed in previous chapters). You can test for nitrates by employing a test kit that is used for freshwater aquariums. You can get started by filling the aquarium to a large extent with water. Then test with the water test kit. You are basically testing to determine if the water is safe for fish which should include a pH on the brink of 7.0, ammonia level of 0 ppm and nitrite level of 0 ppm. You can also ask pet stores to test water samples for you. They usually do it at no extra cost to you. Add 1 cap filled with 100% unadulterated ammonia to the water (to help fish waste simulation). Next, let the system cycle the ammonia throughout the water for a minimum of 3 days. Test the water for ammonia levels (you intend to find a level close to .50 ppm ammonia for the system to start producing nitrites. Keep the ammonia levels stable throughout the method and keep testing every 3-5 days while adding small amounts of ammonia if necessary. After 1-2 weeks (at most) nitrifying bacteria should be present, which at now, you should begin testing for nitrites that are gotten from the nitrification process. The ultimate system fully cycled will have diminutive zero to no ammonia or nitrites and a coffee quantity presence of nitrates. As soon as the water test authenticates the existence of nitrates then the system is proven and prepared for fish and vegetables.

Step 14: Addition of fish and vegetables. There are several factors that you must put into consideration before choosing which fish to start your aquaponics system with. For instance, the temperature will play an enormous influence on what sorts of fish will grow well in certain temperature environments. Also, do you propose to consume the fish since some fish aren't the best for eating while some are best suited for commercial purposes? You can start with any amount of fish you desire depending on the dimension and capacity of your aquarium. But as a beginner, you are advised to begin with nothing above 10 fish preferably a tilapia.

Introduce fish into the aquarium by following the instructions from the pet store. Then let fish stay in transport water to adjust to the temperature inside the aquarium you built for approximately a quarter-hour. Then remove fish out of transport water bag to not mix within the transport water. For a system still within the beginning stages, planting already sprouting vegetables would be suggested however opting for seed remains possible. Now you've got everything needed to start cultivating your own vegetables and raising fish whether for personal consumption, pet, or commercial purpose. There are a lot of other systems designed by many other aquaponics gardeners hooked into personal preference, budget, space and supplies available.

Chapter 5

Aquaponics System Maintenance

If sustainability is your aim, then you have to apply both routine and weekly maintenance activities. Maintenance ensures your aquaponics system thrives successfully. Some of the activities you as an aquaponics farmer needs to carry out at the barest minimum and mostly, weekly include;

- Readjusting the pH if needed
- Harvesting fish if necessary
- Checking water quality
- Checking if there is contamination in the fish tank
- Checking for pests
- Adding vegetables and harvesting vegetables if necessary
- Inspecting if there are plant deficiencies

However, to ensure a smooth maintenance system, an aquaponics farmer needs to keep a maintenance checklist. While it might appear unnecessary, a maintenance checklist will help maintain stability in your system.

What then is an Aquaponics Maintenance Checklist?

Think of it by way of a guide. A guide of what is already done or should be done in an aquaponics system. This can be a daily, weekly, monthly, or routine guide. It is often ticked against tasks that have been done and those that are yet to be done in an aquaponics system.

Aquaponics maintenance checklist can make available a predictable guide for supervision and management of your system, and also provide workforce (if you have any) with a simple timetable to follow when required. Having a checklist helps you maintain focus on the various areas of your setup that need intimate and immediate care. Issues like averting diseases that your fish might face and keep a constant tab on the water pH values are important to preserving a healthy ecology. It is your checklist that ensures you do not miss out of any maintenance routine step for your system.

Why is having an Aquaponics Checklist vital in an aquaponics system?

Perhaps the major reason for an aquaponics checklist is to lessen the frequency of not performing necessary maintenance. Maintaining an accurate record of your aquaponics checklist is also important primarily if you've got a good number of assistants supervising your system. This way, every assistant will have a thought of what they need to do, and maintaining a checklist will greatly reduce unnecessary errors and negligence.

Daily Aquaponics Checklist

The following are some easy and daily routine supervision tasks you'll follow:

1. Feeding the fish: This is the first thing you must tick off on your checklist. Has the fish been fed? It is recommended that you give your fish food twice to thrice each day if manageable. Subsequently check in a quarter-hour if the fish consumed all the feed given to it. If not, remove all the leftover feed within the tank and adjust quantities subsequently to attenuate further food waste. Fish feed is not cheap. Feeding is an imperative part of an aquaponics system. It is imperative to ensure your fish are fed adequately in the least times. Confirm you feed them a minimum of once each day, although two meals

are highly recommended; one within the morning and the other before sunset. It's always great to be physically present when giving your fish their feed because you'll perform health checks and lookup for unusual behavior. However, you will want to think about getting an automatic feeder to assist you if you aren't available.

2. Inspection of fish behavioral changes and appearance change. Fish feeding is the perfect time to watch fish behavior. Check for the level of excitement in the fish when they see food. Are they clumsy towards the food? Before you feed your fish, observe their behavior. In the same manner, observe to see if there is any change after you have fed them. The idea is to note if there are unusual behaviors indicating the presence of ill health or stress.

3. Determine the temperature: Check whether the temperature of the aquarium is the right temperature requirement for your fish. Adjust, if necessary. Temperature provides the right

environment for the species you grow. It is imperative to note that a balanced temperature equals a viable system.

4. Insects: Check your aquaponics system consistently for insects to ensure any insect problems are corrected before they get out of hand. Insects can kill your vegetables. It's also good to recollect that the majority of insects tend to remain in stem sections and under the leaves of vegetables.

5. Assess to see if your pumps (water and air pumps) are functioning properly: You have to check your air and water pump consistently. There are principally two explanations why water and air pumps fail. These being that electrical power is scarce or possibly the water pumps are defective. In either case, you ought to address the difficulty directly because the absence of water pumps will lead to a drop in the amount of oxygen. If you've got defective water pumps, change them immediately Also, consider

connecting standby power supplies or generators just in case of crises or power outage in your area.

6. Scanning for leaks: Your fish tank can start leaking without you knowing. In circumstances where you detect that the movement of water through your aquaponics system is low or you noticed a piling of water at rock bottommost of your system, there are likely leakages somewhere within the tanks or pipes. If not properly taken care of, this might cause water drainage issues and eventually result in both crops and fish not getting adequate nutrients. The moment you encounter this scenario, straightaway mend the noticeable holes or leaks and replenish the water.

7. Assess system water movement: It is important to make sure your system has adequate aeration. Your fish needs aeration to survive; hence the need for a functional water circulation system. Get rid of things (gravel, sand, stone, and clay)

that you notice are causing a blockage in your system as soon as you notice there are blockages.

Now that we have looked at the daily routine checklist. Let's delve into the weekly maintenance checklist.

Weekly Aquaponics Checklist

Conduct water quality test: Examine the amount of your pH, nitrate and nitrite and ammonia and other chemicals before you begin feeding your fish. Kindly note that there exists a difference between nitrate and nitrite and you have to check for their different levels in the water. If there's a drag with the water, it can cause pressure, disease, and even a great death rate of your fish colony.

2. Check to see if there is a need to readjust the pH level of your setup: It is not unusual for the pH of the water to fluctuate due to its environment and temperature. It could be too acidic which is not ideal for your system. A balanced pH level is what your system needs to thrive. It's optimum

that the water remains in its best pH range requirement so that stocks can access all available nutrients. The pH also regulates the health of your fish, the formation of germs within the system, and therefore the capability of your vegetables to soak up nutrients. That's why the pH value of aquaponics plays a significant role in defining whether your aquaponics will work. Check the pH value weekly. The perfect pH for many systems is typically 6.8-7.0, and a few systems can sustain it regularly. The pH level for many systems, however, decreases naturally. Always increase the pH if it drops below 6.5 by adding hydrated potash or lime.

3. Nitrate and ammonia concentration levels: Examine the ammonia levels weekly for any problem. An unexpected increase in the ammonia level, for instance, is a pointer that the system may have a dead fish. The amount should not exceed 0.5 ppm. On the opposite hand, nitrate levels must be checked monthly and must not surpass 150 ppm. Levels that go beyond this are

an indication that the vegetables within the system are consuming less nitrogen within the nitrogen released by the nitrifying bacteria. You can resolve this by growing more vegetables or harvesting some fish.

Gather fish if necessary: The moment you discover that a fish is ready for harvest, waste no time in doing so. In an aquaponics garden, growers must keep a stable stocking concentration of fish. Thus, it's important to reap when essential to take care of this balance.

You can add vegetables and harvest vegetables if essential: You can also add and harvest vegetables during your maintenance. Matured vegetables should be harvested while you quickly plant new crops to maintain the equilibrium of your system.

6. Examine if your vegetable has diseases or infection affecting their growth: Your vegetables can be affected by diseases which can lead them

to suffer retarded growth, or their leaves changing their color or having holes in them. If this occurs, it's likely the vegetables are deficient in vital nutrients. It is recommended that you test the quality of your water to determine if they are ideal for the growth of your vegetable.

Monthly Aquaponics Checklist

1. Fish supply restocking: As hitherto stated, restocking your fish tank is significant to take care of the equilibrium of your system. There should be an equilibrium between the fish in your tank and the quantity of vegetables that you are actively cultivating within the system. Remember it is s symbiotic relationship.

   Assess your fish for diseases: It is pertinent that you assess your fish for disease. Even with preventive estimates, sometimes disease will break out in an aquaponics system. Hence, it is highly recommended that every month, you sample a certain amount of your fish and check for the existence of diseases. This will help you identify early indications of disease and stop it from dispersing in your tank. In fact, assessing your fish for diseases is important because an outbreak of diseases in your fish tank can affect your vegetable's growth.

3. Cleanse all of your purification tools: Remember that ammonia is toxic to your vegetables, hence you need to often do regular cleaning of your filters, clarifiers and biofilters. To avoid blockage, contemplate executing a monthly cleaning of your filters, clarifiers, and biofilters. When something blocks your system's pipes, it's going to end in low effectiveness of your filters and a rise within the fish tank's ammonia concentration level.

Chapter 6

Fixing Common Mistakes in Aquaponics

Since no system is safe proof, it is pertinent to note therefore that there is usually one mistake or the other that can arise in an aquaponic system. The list of mistakes made in an aquaponic system are endless. However, the major mistakes will be treated in-depth.

1. Low water quality: Adding poor or low-quality water into your system poses a challenge to a viable aquaponic system. Maintaining good water quality is extremely important. Water, being the blood of an aquaponic system, transports and supply all necessary nutrients to the vegetables, and it's the medium through which the fish live and survives.

   5 key water parameters to control and monitor:

   - Dissolved oxygen (5 mg/liter)
   - pH (6–7)
   - Temperature level (18–30 °C)

- Concentration of nitrogen (Ammonia, nitrites, nitrates)
- Water alkalinity

In addition, you have to check the chlorine, pH and pathogens/parasites levels before you introduce it in your aquaponics system. It's recommended to check for chlorine with a test kit because chlorine is often deadly to fish. The common test kits make complex water chemistry and management simpler and easier. If you're fixing a fresh system or have a running system and wish to feature water thereto, you'll need to defuse the water by letting it sit for 48-72 hours with the aeration unit running.

Unbalanced fish to water ratio: Adding more fish than a system can handle is detrimental to an aquaponic system. Excess quantity of fish in a tank or introducing numerous amounts of fish in a tank can stress the fish. Another common problem among those starting out is that they often want to grow as many fish as possible and as quickly as possible so they can make sufficient profit from the system. However, they find themselves overcrowding their fish tanks. There is a size density allowed for a fish tank. Overcrowding your fish tank doesn't only end in high nitrate levels and stunted growth for the fish, but also can cause fish loss. It is advisable to keep stocking densities low when starting out with the system. Never overcrowd your fish tanks. With lower stocking density, you'll be able to easily manage your aquaponic system and may avoid any shocks and collapse. The endorsed stocking density is 20 kg/1000 liters, which also leaves room for substantial plant growth.

Wrong choice of growing media: Many people start with wrong media like expanded shale or limestone, ignoring some basic factors. There are numerous aspects of a media you need to contemplate before choosing your growing media. People prefer using media they can get locally and easily, but they don't know what media they ought to use. The best grow media is nutrient-free, with a neutral pH, ready to retain air, ready to retain water and drain quickly so that roots don't get waterlogged. Growing media also allow the colonization of microbial populations. Hydroton is the best ordinarily used growing media due to its good water retention, neutral pH and straightforward to take care of.

4. Using harmful additives to lower the pH: While it is good to lower the pH of your system, it is also poisonous to use harmful additives to lower it. Some impatient people attempt to lower their pH by employing a chemical like hydrochloric acid. Although chemicals will adjust your pH, many of them can potentially harm the well-being of your vegetables and fish. So don't think about using random chemicals for the system. The nitrification process will decrease the pH of your system over time which is safe for your fish too. Vinegar is additionally a slow fix, keep the dosages small.

Planting and growing vegetables unsuitable for an aquaponic system: Early research indicated that leafy vegetables and vegetables like lettuce, chives, etc. did best in an aquaponics system. However, newer research shows that a large proportion of more varieties like tomatoes, cucumbers, peppers, melons, herbs, etc. also do alright in an aquaponics system. Though, some growers plant crops that cannot thrive in an aquaponics system. Crops like yam and cassava among others. You should stay with crops that are already established as fit for an aquaponic system. If you intend to grow non-native vegetables, confirm if they are doing well. Grow seasonal vegetables in their season. You should grow faster-growing vegetables, like salad greens, and vegetables with longer-term crops (eggplant). Remember, vegetables generally do well in aquaponics. If you're growing vegetables that are low-priced in the market, you will be making an enormous mistake that will cost you potential profits. You should also take advantage of yellow peppers, tomatoes, basil and other herbs that support greenhouse. You should not grow any grain crops such as corn and wheat that are only suitable for outdoor fields.

Not having a pest control strategy: If you're keeping the system clean, you'll rarely experience any pest problems. Since you can't use pesticides in aquaponics, you should be vigilant in your pest scouting efforts and quick to react if you see any pest problems. These critters spread and grow in numbers, so better look out for them once you spot them. You can utilize cultural, biological, physical and chemical methods to eradicate pests. If you however find them in your system, there are some ways to remove them. This has been discussed extensively in chapter 3.

Are your vegetables getting enough light?: I hear people say that their vegetables aren't doing well in aquaponics. And most of them try to grow their vegetables indoors or in a shade. One requirement for a plant to grow is light.

Photosynthesis occurs in the presence of light. Vegetables grow by photosynthesis, which needs light of a minimum of 30,000 Lux. So whether you are growing your system indoor or outdoor, your vegetables must have access to adequate light. If you've got no access to sunlight, use LED grow lights. If you've got a little setup, these cheap and little LED grow lights will work fine. Use light meters to watch if your vegetables are getting enough light for their rapid growth.

8. Using transparent or translucent fishtanks: Algae love water and lightweight, and they will grow everywhere they get nutrients. They're going to grow so well that they will clog the smaller conduits, like the tubes. This is exactly what happens when you utilize a transparent fish tank. Algae will leach to it. Using opaque material, like PVC, prevents any algae issue in aquaponics because it insulates the sunshine that promotes algae growth. If your system is exposed to sunlight, cover the aquarium and put the floats

into the grow bed. Blocking out the sun kills off the algae that are removed by the filter, in turn.

Do your fish have a regular feeding pattern?: Most people are reckless in feeding the fish. Fish need an adequate amount of food at intervals. You should study your fish and know the optimal time to feed them. Fish when being underfed can cause low fertilizer nitrate levels. You can utilize a fish automatic feeder to automate this job.

10. Not checking the ammonia/nitrite/nitrate levels: Are your fish dying? The amount of unionized ammonia or nitrite may be too high within the system, and you would like to extend the biofiltration. High levels of ammonia are mainly caused by overfeeding the fish or low aerobic activity of nitrifying bacteria. This is caused by tons of factors like unstable pH, and temperature. Fish produces ammonia and discharges it

through their gills. Ammonia is extremely toxic and can eventually kill the fish. Hence, there's a requirement to dilute, remove or change the ammonia present within the water of the aquarium. You ought to test the water a minimum of once every week using test kits to work out the ammonia concentrations within the aquarium.

The end... almost!

Hey! We've made it to the final chapter of this book, and I hope you've enjoyed it so far.

If you have not done so yet, I would be incredibly thankful if you could take just a minute to leave a quick review on Amazon

Reviews are not easy to come by, and as an independent author with a little marketing budget, I rely on you, my readers, to leave a short review on Amazon.

Even if it is just a sentence or two!

Customer Reviews

★★★★★ 2
5.0 out of 5 stars ▾

5 star		100%
4 star		0%
3 star		0%
2 star		0%
1 star		0%

Share your thoughts with other customers

Write a customer review

See all verified purchase reviews ▸

So if you really enjoyed this book, please...

>> Click here to leave a brief review on Amazon.

I truly appreciate your effort to leave your review, as it truly makes a huge difference.

Chapter 7

Aquaponics Frequently Asked Questions

1. **What sorts of fish are often utilized in an aquaponics system?**

   The ideal fish for an aquaponic system is tilapia. Tilapia is the commonest due to its hardiness and its ability to tolerate a good sort of water quality conditions and survive in a high-density environment. One word of caution though, you will have to consult your local regulation body for fish and other aquatic programs if they are regulations on the type of fish you are allowed to grow. Many nations have very strict regulations that have to be followed.

2. **What percentage fish will be useful to cultivate my vegetables?**

The percentage of fish in your system affects the development of your plant whether directly or indirectly. Your vegetables feed on your fish. The number of fish you have to nourish your vegetables will again depend upon the size of your fish and how you feed them every day. It has been proven by a study that if you employ sixty-hundred grams of fish food every day, you'll be ready to grow one square meter of vegetables during a raft aquaponics bed.

3. What kind of feed would be suitable to feed my fish?

The role of feeding in an aquaponic system cannot be undermined. If you'll be raising your fish for food production, you'll want to feed your fish a specific commercial fish food. If you won't be using your fish for food production you'll make your own fish food or use things like Duckweed, pistia, worms, or the other sort of similar live feed.

4. What kind of containers is best for my system?

While translucent containers work, they are not the perfect container for an aquaponic system. It's imperative that you simply use only food-grade plastic containers. Anything could transfer chemicals into your water system and cause damage to you and your fish. You can lose a lot of money just by buying the wrong container.

5. **Is greenhouse necessary?**

Using a greenhouse housing system is dependent on your country and area of residence. In fact, you'll always use a greenhouse if you reside in a really cold region. A greenhouse is ideal for a year-round aquaponics system because it is viable for speedy healthy growth and protection of your system. Additionally, if you reside in an environment that is harsh then a greenhouse is suitable and helpful in protecting your vegetables from breaking winds and heavy rains.

6. **Will my aquaponics systems thrive indoors?**

The answer is yes. A lot of persons maintain and control their system inside their homes, but you have got to confirm if your system will thrive for indoor situations like man-made lighting, etc. You have to have the necessary conditions for an indoor system.

7. **Will my aquaponic system consume a high voltage of power?**

The quantity of power consumption of your system will depend on its complexity and size. Although normally, water pumps consume about 25 watts of energy which is considered okay.

8. **Can I feed animal waste to my fish?**

No. Animal waste could contain e-coli and will pollute your system. It is either you prepare your fish feed or you buy fish feed for the health of your fish.

9. **What proportion of time should I spend every day looking after my system?**

Daily maintenance is merely about five minutes, which mainly consists of giving food to the fish and checking to ensure all of your water pumps are functioning properly.

10. Do I ever need to pour out water?

No, you don't necessarily have to. However, little proportion of the water is going to be lost thanks to evaporation. Therefore you will just have to add water in between. The essence of adding water is to help your system remain balanced.

11. Will I be able to grow shrimp or crayfish in my system?

We don't endorse it because shrimps are predators. They will eat your vegetables if care is not taken.

12. Can I use a solar power system to run my setup?

Yes, you can

13. Can I use insecticides and other artificial chemicals to destroy the pests?

No. Chemical pesticides will destroy your fish. There are other methods of controlling pest in your system instead of chemical pesticides. You can kill them manually as soon as you notice their invasion. Ladybug are used by some farmers. Otherwise, you could combine 1 / 4 cup of molasses, one teaspoon of chili powder and about four to 5 drops of dish soap in a gallon of water and spray it on your vegetable.

Conclusion

Aquaponics is a system that allows for easy and enhanced maintenance of the hydroponic planting system. It preserves this planting method and enhances its efficacy. It also improves fish production by helping to keep their habitat clear and healthy for them to dwell in. It is a system that is cost-effective and highly profitable to set up and run if you carefully follow through with the guidelines discussed in the pages of this book. Aquaponics is the future of sustainable farming as it helps to preserve the fertility of the soil due to the effects soil farming has on the soil, thus making it more porous to environmental issues like pollution.

Aquaponics system can be managed by anyone with training or fundamental knowledge of the subject, which is what this book hopefully has helped you gain. With the knowledge encapsulated herein from start to finish on how to successfully run an aquaponics system, you are surely capable to set up your own DIY system to raise plants and fish.

So then, I wish you all the best!

www.ingramcontent.com/pod-product-compliance
Lightning Source LLC
Chambersburg PA
CBHW071435210326

41597CB00020B/3797